# the ninja star

Art of Shuriken Jutsu

# the ninja star

## Art of Shuriken Jutsu

Katsumi Toda

Published by

**dragon books**

*Acknowledgements*

Publisher/David Chambers
Lesley Anne Copp-Taylor/Editor
Malcolm Copp-Taylor/Line drawings & photography
Garry O'Keefe B.A. (Hons)/Layout & design
Haruko Chambers ⎫
Yumiko Yokoma ⎬ Special Translation

L.C.C.C. No. 84-71966
ISBN 0 946062 10 2

Printed and bound in Great Britain by
Anchor Brendon Ltd, Tiptree, Essex
First Published November 1984
Third impression November 1986

Distributed in the USA
by Dragon Publishing Corporation Thousand Oaks California

*"HOWEVER SERIOUSLY SYSTEMS AND MEASURES MAY
BE DISCUSSED,
THEY CANNOT BE PUT INTO PRACTICE UNLESS THERE IS
THE RIGHT MAN TO DO IT.
THERE IS NO DEED WITHOUT A DOER.
TO HAVE THE RIGHT MAN IS THE GREATEST BLESSING.
ONE MUST AIM AT BEING THAT MAN."*

*SAIGO TAKAMORI
1827-1877*

# Contents

Chapter One
**Historical Background** 1

Chapter Two
**The Ninja** 8

Chapter Three
**The Samurai** 13

Chapter Four
**Shuriken Jutsu** 17

Chapter Five
**Shuriken and Shaken** 20

Chapter Six
**Preparation for Practice** 25

Chapter Seven
**Stances and Grips** 30

Chapter Eight
**Throwing and Targets** 36

Chapter Nine
**Mind Control and Kata Form** 43

Chapter Ten
**Special Drills and "Secret Methods"** 55

# Message from the Author

## TO ALL MY UNKNOWN FRIENDS:–

*It is with great pleasure I pick up my brush and set before you this dissertation upon the arts which have become known as "Shuriken Jutsu".*

*To those of you who enjoyed the saga of Tomokatsu Ryu in "Shadow of the Ninja", "Revenge of the Shogun's Ninja" and "Ninja Death Vow", and have expressed your kindness in letters care of my publishers Dragon Books, I ask that this be accepted as my gratitude for such pleasant sentiment. I regret, that so great has the response been, it is impossible for me to adequately answer all of you...... my friends.*

*I will, however, give a brief outline of my experiences. I have studied deeply the ways of sword and spear, these many years. The blood of warriors and priests flows in my veins – on my paternal line, we trace back seventeen generations in loyal service. I am an advocate of pen and sword – neither one is more important and neither should be neglected.*

*Daily practice is the fountainhead of my philosophy. A day should not start without practice (Renshu. Translator's Note). I am fortunate in that I live almost one hundred and fifty miles from the noise of the great capital. I have, what may be described as a rural estate. The land is flat and fertile. We face south with the sea some four miles west of us, and mountains behind. It is in this setting that my written work is done. I have an area of my garden which suffices as my dojo. In rain, sun, wind or snow it is here that I greet the dawn. I have had my illustrator draw some of my training implements, which I include in the latter part of this book. I hope it may be of help to you.*

*Although my background is one of consanguinous classicism, I can appreciate the interest and desire of the young to create the new and vibrant forms. It is not within me to condone or condemn such action, provided it is performed with a light spirit and eager heart.*

*To the many unknown friends
I offer a heartfelt welcome.
I remain, yours sincerely,*

**Toda Katsumi**

*I would like to take this opportunity to publicly thank my special translators; Mrs. Haruko Chambers and also Miss Yumiko Yokoma for their help in the production of this book.*

# WARNING

*The arts as described in this book are totally for self-perfection. Persons attempting to use the techniques contained in this book for evil purpose, have completely misunderstood the underlying principle of this book, which rests upon the principle of:-*

*ONORE O SAMERU – To control the self. I would add this advice – First know yourself, then begin to know others. We are all pupils at heart.*

**Toda Katsumi**

*View at Futami-Ga-Ura on Shima Peninsula. The Myoto Iwa "Wife and Husband" rocks Izanagi and Izanami – mythical founders of Japan.*

# CHAPTER ONE

# Historical Background

KWANNON

*"In the land of Yamato*
*The mountains cluster*
*But the best of all mountains*
*Is Kagu, dropped from heaven*
*I climbed, and stood, and viewed my lands"*

*– Emperor Johei (Nara period 794 A.D.)*

"The early dawn mist hangs low over the dry moat of the great castle, huge stones joined by giant's hands, rest heavily; rising to the parapet, which is topped by the mighty castle keep. the plaintive cries of the meadow shrike drift across the cold air. High upon the parapet a solitary samurai guard stops in his lonely vigil, bringing the fuse of his Tanegashima matchlock gun to his lips, he blows the ember into a bright red core. From far below, dark eyes have seen his action. With a hiss, the six-pointed death star snakes its way in a deadly arc. The samurai guard drops silently in death. From all around, the black-clad ninja warriors rise to storm the castle......."

Such is the popular myth of Shuriken... the ninja star. The truth as always, is far stranger. Through the pages of this book, I hope to show that the arts of Shuriken Jutsu are indeed wide and varied. But before we embark upon our voyage of discovery, there are certain facts it is worthwhile to pause and ponder. As the years grow long, the remembrance of what was true and not so true becomes blurred. Thus is history, a product of many men's interpretation. What may seem mundane to one historian, may have a deep significance for another.

The modern disciplines such as Kendō, Karate-Dō, Judō and Aikidō are essentially products of the modern era. Their philosophy and goals are the product of a modern educative system and owe more at their heart to self-perfection, than to self-protection.

Whilst it may ruffle the feathers of many ardent students, I must state that these modern systems have little or nothing to do with battlefield-proven methods of the classical warriors – more popularly referred to as the "Samurai". Whether you call it Dō, Bushidō, or

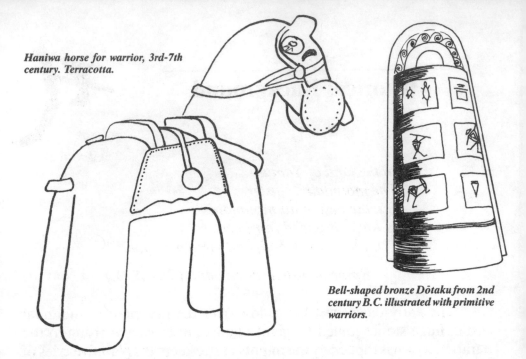

Haniwa horse for warrior, 3rd-7th century. Terracotta.

Bell-shaped bronze Dōtaku from 2nd century B.C. illustrated with primitive warriors.

Shinkan; the developers of these fine systems were primarily educators; catering, as has been witnessed in the post-war period, to a mass market. There is a serious misunderstanding amongst students, even in Japan, of the heart of the warrior ethos.

As the late Donn F. Draeger, was at pains to point out to Western practitioners; the Kyu-Dan system of awarding coloured belts and black belts as proof of ability and progress, has no part in the classical martial arts of Japan: it's concept is, as I stated, one of an educative system for a mass market.

The classical Ryu are by nature, comparatively small in individual size, though there have been as many as 10,000 separate ryu; for although small in size, their heart is that of Japan.

The essentials for entry to a classical Ryu are as many and varied as the Ryu themselves. But it is worthwhile to note that in general great emphasis is laid upon a sound moral character. At their heart, they tend to be composed of right-wing conservative persons, who set great store upon resolute acceptance of authority in its many forms.

Change, purely for its own sake, has no interest for these persons. As the central core of most of the Ryu lies in an unproven feudal past, it is difficult in many cases to provide an accurate founding date. But as many of the founders claimed Divine Inspiration and in some cases, intervention in the evolution of their styles. The oft repeated arguments for change in the face of society's changing values, is met with the simple

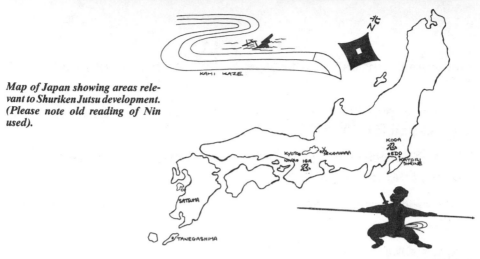

*Map of Japan showing areas rele-vant to Shuriken Jutsu development. (Please note old reading of Nin used).*

rejoinder, "why practice the mundane, when you have the Divine at hand".

This belief in the underlying divinity of their ways is not mere lip-service, far from it. How often has one entered a dojo, to see the Kami-dana and Kamiza bedecked with trophies and photographs of the latest champions. I am deeply concerned that this should occur in Japan, which should set the standard as the fountain-head of traditional values. It is not surprising that outside of Japan, the Kami-dana be a photo-graph of the instructor. This is, however, a detrimental practice and should be discouraged. When one makes a bow to Kamiza, it is a bow to the very heart of nature, not to one man; no matter how great or illustrious a past he may have.

A word of advice to those of you who do not have an instructor and are searching for one. Please note, any person purporting to teach classical Shuriken Jutsu as an art by itself; away from the code of ethics embodied in a classical ryu; must of necessity, be treated with suspicion.

Even in Japan, there are few bona-fide instructors of Shuriken Jutsu in the classical form. If this is the state in Japan itself, what can it be outside? Before the war, there was little problem, here I speak of the old Kyoto Butokukai, which at least gave a standard by which others could be judged.

To return to our studies, it is necessary for you to have a little back-ground knowledge of our historical eras.

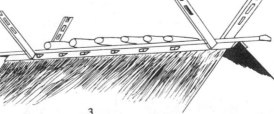

*Roof at Ise, the birthplace of Japan.*

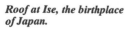

# THE ERAS

**The Yamato period** – Stone Age to the introduction of Buddhism in A.D. 552.

**The Asuka period** – from the introduction of Buddhism in A.D. 552 to the Taika Reformation in 645 A.D.

**The Nara period** – from the Taika Reformation in 645 A.D. to the removal of the capital from Nara to Kyoto in 794 A.D.

**The Heian period** – from the removal of the capital from Nara to Kyoto in 794 A.D. to the fall of the Taira family in 1185.

**The Kamakura period** – from the fall of the Taira family in 1185, to the Emperor Godaigo's transference to Yoshino in 1336

**The Namboku period** – from the Emperor Godaigo's transference to Yoshino in 1336, to the unification of the Southern and Northern dynasties in 1392.

**The Muromachi period** – from the unification of the Southern and Northern dynasties in 1392 to the fall of the Ashikaga Shogunate in 1573.

**The Azuchi-Momoyama period** – from the fall of the Ashikaga family in 1573, to the fall of the Toyotomi family in 1615.

**The Edo period** – from the fall of the Toyotomi family in 1615 to the Meiji restoration in 1868.

**The Tokyo (Gendai) period** – from the Imperial restoration in 1868 to the present.

## A SIMPLIFIED LIST OF RYU:–

A simplified glossary of the important classical Ryu of Japan, who enshrined the Samurai spirit in its heart. Many no longer exist as practising groups, but their spirit remains with us.

> *"Is it a shower of rain?*
> *I thought as I listened*
> *From my bed, just awake*
> *But it was falling leaves*
> *Which could not stand the wind."*
> *Priest Saigyō, Kamakura period*
> *(1185-      )*

# THE RYU

(Note that dates refer to approximate founding and should be viewed in such a light.)

| | | |
|---|---|---|
| AOKI-RYU | 1634 | KAN-EI |
| ARAKI-RYU | 1638(?) | KAN-EI |
| ARIMA-RYU | 1480(?) | BUNMEI |
| BOKUDEN-RYU (See also Kashima Shinto Ryu) | | |
| | 1520 | TAI-EI |
| CHUJO-RYU | 1394 | OEI (Appochryphal) |
| CHUYA-HA-ITTO-RYU | 1704 | HO-EI |
| ENMYO-RYU | 1650 | KEI-AN |
| GAN-RYU | 1600 | KEI-CHO (Appochryphal) |
| HASEGAWA-RYU | 1573(?) | ITEN-SHO |
| HOKUSHIN-ITTO-RYU | 1830 | TEN-PŌ |
| HORIUCHI-RYU – synthesised | 1854 | ANSEI from KANSEI (1789) |
| ICHIU-RYU | 1480(?) | BUN-MEI |
| ITTO-RYU | 1704 | HO-EI |

**See also**    HOKUSHIN ITTO RYU
ITO-HA ITTO RYU
KAJI-HA ITTO RYU
KOGEN ITTO RYU
TENSHIN ITTO RYU
YUISHIN ITTO RYU

| | | |
|---|---|---|
| JIGEN-RYU | 1600(?) | KEI-CHO |
| KAGE-RYU | 1532 | TENMON (Appochryphal) |
| KANEMAKI RYU | 1596 | KEI-CHO |

| | | |
|---|---|---|
| KASHIMA SHINTO RYU | 1520 | TAI-EI |
| KATORI SHINTO RYU | 1477 | BUN-MEI |
| A.K.A. TENSHIN SHODEN KATORI SHINTO RYU | | |
| KURAMA HACHI RYU | Date vague | Certainly active in EARLY TOKUGAWA and before |
| KYOSHIN MEICHI RYU | Date vague | EARLY TOKUGAWA |
| MANIWA NEN RYU | 1532 | Believed origin 1200 TENMON & SHOJI |
| MIJIN RYU | 1596 | KEI-CHŌ |
| MUSASHI RYU | 1645+ | KEI-AN |
| (See ENMYO RYU & NITEN ICHI RYU) | | |
| MUSO JIKIDEN EISHIN RYU | 1716(?) | KYO-HO |
| NEN RYU | 1532 | TEN-MON |
| NIKAIDO-RYU | 1596(?) | KEI-CHO |
| NITEN ICHI RYU | 1645(?) | KEI-AN |
| OMORI RYU | 1716(?) | KYO-HO |
| OKUYAMA RYU | 1573(?) | TEN-SHŌ |
| ONNO HA ITTO RYU | 1661 | KAN-BUN |
| SHINKAGE RYU | 1570 | GEN-KI |
| SHINKAN RYU | 1573 | TEN-SHŌ |

| | | |
|---|---|---|
| SHINKATATO RYU | 1688 | GENROKU |
| SHINTO RYU<br>See KATORI SHINTO RYU | 1477 | BUN-MEI |
| SHINTO MUNEN RYU | 1751 | HŌRYAKU |
| TAKEMORI RYU | Date vague | MID<br>TOKUGAWA |
| TAKENOUCHI RYU | 1538 | TEN-MON |
| TAKENOUCHI HANGAN<br>RYU | 1715 | SHOTOKU |
| TETSU JIN RYU | 1645(?) | KEI-AN |
| TODA RYU } Believed<br>TOMITA RYU } similar origin | 1564(?) | EIROKU |
| YAGYU RYU | 1647(?) | SHŌ-HO |

This is by necessity, a simplified list, my sincere apologies to any Ryu I have not included.

# CHAPTER TWO

# The Ninja

*"O – the dark shadows*
*Let them clothe me in night"*

*Anon*

There can be no more emotive an image than that of the black-clad ninja warrior noiselessly scaling battlements, an elusive phantom, master of the occult and the divinatory arts. The truth is once again, far stranger.

In the mythical past, the ninja were said to have evolved from the long-nosed winged goblins known as Tengu. Their secret method was in the form of a synthetic science known as Omyodo, based upon Chinese principles of divination and astrology. It has been suggested by certain scholars, that the Yamabushi warrior monks (not to be confused with the peaceful mountain sleepers) were the factual fore-runners of the ninja; foraying forth, as it were, from Mount Hiei.

By 1185, with the dawning of the Kamakura period, the stage was set for the tremendous upsurge that would culminate in the autonomous ninja clans.

The tumultuous fall of the Taira clan, coupled with the weakness of the Kyoto government engendered much of the populace to seek

*Tengu demon, the mythical fore-runner of the Ninja.*

*Yamabushi monk from Mount Hei, armed with naginata and wearing full battle armour under his robes.*

*Ninja!*

methods of self-protection, both physical and spiritual. As a direct result, numerous training camps sprang up purporting to give warriors an "edge" with the rudiments of tactics.

Legend has it that Yoshitsune, who is likened by Western writers to Robin Hood, was trained by sword masters of the Kurama Hachi-Ryu in the wildlands of Mount Kurama, finally perfecting his own semininja form, known as Yoshitsune Ryu. It appears to have been a rather direct form, involving the "pre-emptive strike", rather than the introspective forms of espionage as a form of political control. It was not free however, of the constraints laid upon it by the family system.

The transition from Heian to Kamakura period saw well over

*Heavyweight Shuriken.*

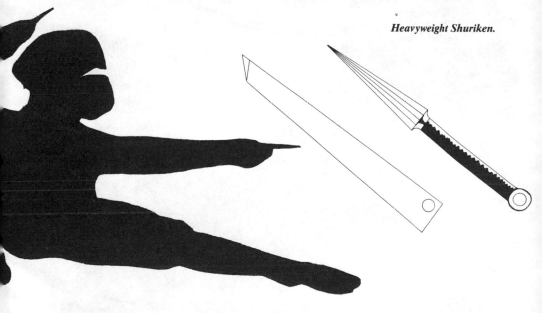

twenty ninja clans and here I refer to "clan" as a collection of like minded people, not necessarily a blood line. However the arts of the ninja were still considered worthy of inclusion as only a small part of a Ryu's extensive curriculum, as is evidenced by the Ryu to which belongs the longest unbroken line – the Katori Shinto Ryu (founded 1477 by Iizasa Choisai Ienao), truly Tenshin Shoden.

It was not really until Lords Hattori and Oe of Iga Province set the scene for the ninja clan as an entity independent of the Ryu that the age of the full-time ninja dawned. Coincidentally, it was also the age of the great battles and of intrigue – perhaps not such a coincidence after all!

Another name which cuts its way out of the pages of history to found a mighty ninja clan was Kusonoki Masahige. It was reputed that he held Kyoto in the grip of an iron ring of 48 highly skilled ninja agents.

By the time of the sixteenth century, the provinces of Iga and Koga were the breeding ground of colonies of ninja, many owing a degree of

*On the battlefield – a Ninja, his sword broken defends himself with a Shuriken.*

allegiance to one clan – but a greater amount for hire to the highest bidder.

Lord Toda Takatora of the Iga province was an example of a ninja entrepreneur. He is reputed to have recruited likely candidates from surrounding provinces and trained them in the esoteric mysteries at Hakuho Castle.

When the warring clans had finally been subdued by the might of the Tokugawa forces, the Shogun's peace of two hundred and fifty years, was policed by the Bakufu government whose Metsuke utilised the skills laid down by the Shogun Tokugawa Ieyasu's ninja, Hattori Hanzo.

There are various unconfirmed stories of the activities of ninja-like forces during the conflict which saw the restoration of Imperial Rule in 1868, on these I cannot comment.

The arts of Shuriken Jutsu were important to the feudal ninja, but really they were not, as film-makers would have us believe, their sole preserve. Indeed the Samurai were equally at home with the constructive use of Shuriken and Shaken.

## FAMOUS NINJA
### *(Japanese form, family name first)*

| | |
|---|---|
| Momochi Sandayu | Kasumi Rosuke |
| Hattori Hanzo | "Sarutobi" Sasuke |
| Fuji-Bayashi Nagato | Ukifune Jinnai |
| Ishikawa Goemon | Saji Gorobei |
| Yamada Yaemon | |

*All active late 16th, early 17th century.*

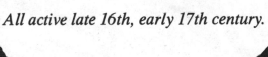

*Foot soldier wearing Dōmaru after Moko Shurai Ekotoba Emaki – late 13th century.*

# CHAPTER THREE

# The Samurai

FUDŌ

*"Pride of a thousand years, stainless and terrible
Flood water powerful and plum blossom fair"*

**Captain Raymond Johnes**
*English soldier and scholar*

The finest flower of the nation is said truly to be that of the Samurai. The naming of this ultimate warrior comes simply from a Japanese word meaning, "to serve", for service to a Lord was the cornerstone of life. It was said that to serve was life and to live was service. It is generally accepted that the term, "Samurai", came into being between the ninth and eleventh centuries A.D.

By the twelfth century A.D., the Samurai had become a mighty force to be reckoned with. The conflict between the rival clans of Minamoto (Genji) and Taira (Heike) escalated in 1180 with the Gempei war which raged for five years. At the end of it, the Heike were scattered as dust before the wind and the Samurai warlord, Minamoto Yoritomo, became more powerful than the Emperor himself. It was Yoritomo who coined the title which has become synonymous with Samurai – that of "Shōgun" – literally commander-in-chief.

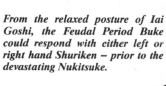

*From the front, the Feudal Buke is peacefully standing, but from behind his Shuriken lie palmed ready for a pre-emptive strike.*

*From the relaxed posture of Iai Goshi, the Feudal Period Buke could respond with either left or right hand Shuriken – prior to the devastating Nukitsuke.*

*Tosei Gusoku – late 16th century. Black areas denote weak points where Yoroi Toshi or Shuriken could be used.*

A century later, the first great test of Samurai courage came with Kublai Khan's attempt in 1274 to invade Japan – the attempt was thwarted. In 1281, with a force, some ten times greater, Kublai Khan sent his Mongol hordes across the sea, to take Japan by storm but storm took Kublai Khan – a great wind sprang up and the sea rose like a mountain. The divine wind, the Kamikaze, blew. It would be six centuries before any nation would attempt to invade Japan again.

By the fourteenth century, the rule of Samurai General, was shaken when the Emperor regained power. This was short-lived, as the Ashi Kaga proclaimed themselves Shōgun, but they were weak and by 1467 blood flowed in the streets of Kyoto – the terrible Onin war had begun. It heralded the end of the albeit tenuously unified, nation. Japan became a collection of tiny kingdoms ruled over by despotic Daimyo. This lasted for over a hundred years and became known as Sengoku Jidai – the Age of War. It was here that the rudiments of the combat systems started to evolve. The icy cold blade of the Samurai ruled without question. By 1543 however, the "round eyes", the Portuguese, brought firearms to Japan – the days of separatism were numbered.

A Samurai, Oda Nobunaga, grasped the nettle firmly – with the aid of the Tanegashima (firearms) and also using ninja, he overcame his rivals with ease. Unfortunately he was murdered in 1583. He was succeeded by Toyotomi Hideyoshi, who managed to continue the unification of Japan. He died in 1598, leaving a young son as heir. The country was ruled by regents, Samurai clan leaders, greedy for the power which rested on such young shoulders.

Samurai chose sides, by the dawn of October 21st 1600, the protagonists faced each other in the misty rain near the village named Sekigahara. What took place, was the largest battle on Japanese soil. By its ending, following plot and counter plot, General Tokugawa Ieyasu emerged victorious. But the victory was greater than any could have imagined. With the institution of the Tokugawa Shogunate, Samurai life became ordered and structured as never before and foreign trade ceased almost entirely. It was, in reality, 250 years of comparative peace. In this atmosphere, many of the arts received a new breath of life – none more so than Shuriken Jutsu. Although large-scale conflicts were out of the question, Samurai still bore the symbol of their rank – the razor-sharp Dai-Sho pair of swords, and disagreements culminated in the death of one or both parties. It was felt that to have "something up one's sleeve" was a definite advantage.

When American and European ships appeared in Japanese waters in the mid-nineteenth century the end of the Samurai as a class was close on the horizon. The Shogunate, in decline, was easily overthrown and the Emperor Meiji restored to power. But the Samurai as an ideal lived on.

## FAMOUS SAMURAI

| | |
|---|---|
| Minamoto Yoritomo | Tokugawa Ieyasu |
| Kusonoki Masahige | Oishi Kuranosuke |
| Oda Nobunaga | Yamaoka Tesshu |
| Takeda Shingen | Yamamoto Isoroku |
| Toyotomi Hideyoshi | |

*A Samurai defends himself with Shuriken.*

# CHAPTER FOUR

# Shuriken Jutsu

MARISHI TEN

*"The gap between my feelings and my skill was
so great — I wonder I went on"*

*Sir John Bettjeman,
English poet laureate, d.1984*

The arts of Shuriken Jutsu as we encounter them today, must necessarily split into three separate entities. They have little in common with each other, save that an object is thrown. Their underlying principles and philosophies are equally dissimilar.

**The three forms are:**   The Classical
The Modern Cognate
The Modern Synthetic.

**The Classical**
In recent years, a few Ryu have allowed non-Japanese to enter on a trial basis. There is no set time limit to this "test" period but it would certainly be expressed in years of regular study, the absolute minimum being two, three hour sessions in the ryu every week for three years. It is generally felt that only over this period of time, spent in continuous study in Japan, can the motivation and fitness for in-depth study of the Ryu's curriculum be assessed. As I stated elsewhere in this book, there will be no "black belt", advancement being expressed in personal achievement, culminating perhaps in the granting of the "Menkyo" — literally, the licence of excellence.

I pause here to mention one of the foremost Western practitioners, who has made the pursuit of excellence his goal. Sadly he is no longer with us. Donn F. Draeger, 1920-1982, who attained the Kyoshi of Tenshin Shoden Katori Shinto Ryu. It is worthwhile at this point to pause and remember him, for he was responsible in a great way, for the dissemination of the true martial arts and ways of Japan. We are the poorer for his passing.

*Example of extremely heavyweight Shuriken.*

## The Modern Cognate

In essence, the Modern Cognate form of Shuriken Jutsu is practised as an adjunct to serious training in one of the modern Do or Jutsu forms, such as Ju-Jutsu, Judo, Aiki-do, Kendo or Karate-Do. Practictioners tend to adopt the stance names from their own main systems and their practice progresses accordingly. The practice in this form is a highly disciplined method, it makes no claim to being a spiritually and combat effective whole as the classical form, but its interest lies in the development of skill and the sense of achievement in the bringing together of the reflexes between hand and eye.

## The Modern Synthetic

The Modern Synthetic is the result of movements both in America and Europe to nationalize martial arts and ways. For those who follow a classical Ryu, or a modern cognate form, such a concept is inane and without need for further comment. But those who assiduously follow the modern synthetic disciplines cannot help but be excited by the prospect of such turmoil. One only has to glance back to the late nineteenth century with Funakoshi Gichin Dai Sensei and the nationalization of Okinawan Karate-Do, or to the immediate post-war period with Oyama Masutatsu's creation and development of the awesome Kyokushin Kai form of Karate.

*No matter what age, the form remains the same.*

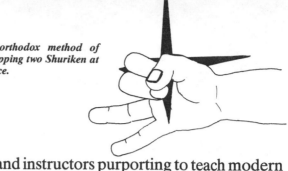

*Unorthodox method of gripping two Shuriken at once.*

The state of instruction and instructors purporting to teach modern synthetic is again as woeful as the classical and modern cognate. There are many fine instructors who are pursuing the development of modern synthetic with the laudable sentiment and wish for excellence. A word of warning:– many modern synthetic instructors, both Japanese and non-Japanese, make fatuous claims concerning the deadly effectiveness of their techniques. Such claims are not the stuff of the true martial artist. They are usually said to impress gullible and usually younger persons. From a logical point of view, such instructors are by derivation, claiming to be law breakers; even murderers. Martial arts and ways have been plagued with this sort of dangerous nonsense for far too long; let us strive to rise above it.

It is a worrying phenomenon of recent years, particularly on the non-Japanese scene, for so-called instructors to claim origins in special forces, green berets, CQB squads (close quarter battle), SAS and SBS and GSG9 (West Germany), or even mercenary covert action groups. The best advice is to leave this pathetic person to his or her dreams and not allow them to egrandize their own ego at the expense of your pocket.

*A modern exponent of Iai Jutsu supplementing Shuriken Jutsu practice.*

# CHAPTER FIVE

# Shuriken and Shaken

BISHAMONTEN

*"While the leaves of the bamboo rustle*
*On a cold and frosty night"*

*Frontier guard. Nara period*

Contrary to popular myth, the Shuriken is not a star shape, this is more correctly termed Shaken. However, in common parlance, "Shuri-kenjutsu" is used as an all-encompassing term.

The development of Shuriken and Shaken is one of great debate amongst students of both Shurikenjutsu and hoplology (academic study of weapons systems). With the development of the Japanese sword, the development is clear and extremely well-documented with over eight centuries of written work.

Being the soul of the Samurai and a part of the Imperial regalia (Mikusa no Kandakara) mirror, comma-shaped jewel, and sword, the sword was always in the forefront of the Japanese mind. Shuriken and Shaken however, were weapons of stealth and surprise – essentially, "throw away" (no pun intended). They were, in the main, of fairly crude manufacture. Unlike the sword, which has passed down the centuries in fine condition, antique Shuriken and Shaken have not fared so well. By their very nature, once thrown in the heat of battle, or on a moonless battlement, they were rarely recovered – the reason for their existence was one of escape or evasion.

The Japanese sword developed into the awesome single-edged curved weapon we see today, from the double-edged Ken of the Yamato Period pre A.D. 552. But what of the Shuriken? The following is expressed as a personal opinion and I would welcome informed academic criticism.

In the repository at Shosoin Nara are nearly seventy small sheaths or pouches containing a veritable arsenal of small knives. Indeed, the late Sato Kan'Ichi (Kanzan), Executive director of the Nippon Bijutsu Token Hozon Kyokai and vice-director of the Token Hakubutsukan (Society for the Preservation of Japanese Art Swords and Sword Museum, respectively – Translator's note) is dubious as to the exact use

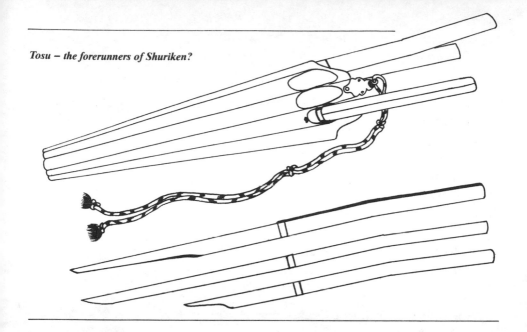

*Tosu – the forerunners of Shuriken?*

of these very early edged tools. Known as, "Tosu", they are implicated in an attempt on the life of Emperor Suinin (26 B.C. according to tradition). The chronicle "Nihon Shoki", states that this occurred in the fourth year of his reign.

The Taihoryo of 701 A.D. mentions them as an object of rank – thus firmly lodging them with the Buke forerunners of the Samurai and ninja, and not as with the development of nunchaku and tonfa, which clearly have an agrarian parentage.

Later on in the development of small weapons, we see the growth of Mitokoromono, literally things for three places, i.e. Kozuka, Kogai, and peculiarly to the Samurai of Higo Province, the uma bari, a horse needle. There are many opinions concerning the use of these items. It is my opinion that both Kozuka, Kogai, wari Kogai (split Kogai) and uma bari, could have been used as rudimentary Shuriken. Their extreme relative lightness preclude their serious use as "stopping" weapons, they would have been admirable as objects of distraction and surprise. As a point of interest, a rather fanciful theory is that the Kozuka could have been used as a "calling card", which would be left in the body of a despatched enemy, thus enabling the local magistrate to tally his records.

Whatever the case, Shuriken and Shaken were designed for surprise and as such, would be secreted about the body or at strategic points in the house or garden – thus affording time for a seemingly unarmed Samurai to reach his sword. In fact we have an interesting example which has come into everyday use in common speech – Years

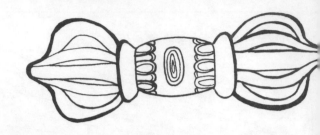

*Vajra Thunderbolt.*

ago Samurai might wear a Futokoro Katana, traditionally a sharp surprise-weapon or knife similar to the Aikuchi tanto, which would be hidden in the folds of inner and outer Kimono. Its modern use is when one wishes to accomplish a great task and cannot go on further by one's own strength – one must seek a hidden ally or advisor – such a person is known as "Futokoro Katana".

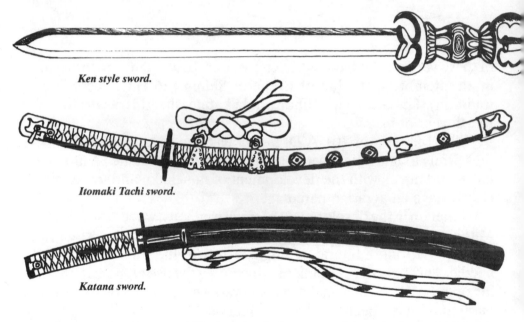

*Ken style sword.*

*Itomaki Tachi sword.*

*Katana sword.*

To set Shuriken and Shaken in their correct historical place we must stretch back as far as the first pre-historic man discovering that a thrown object could bring down an animal or settle a boundary dispute – the rudiments of an instinctive aerodynamics. The classical Ryu all had their own peculiar shapes and sizes, some front-weighted, some back-weighted, some single-ended and some double-ended. Their shapes are lost in the mists of time, save for the personal preference of a particular Samurai or ninja. We must be aware that Shuriken is not really a single implement, but more a collection of implements, Tonki, with different shapes for different jobs.

It is a gross misconception to assume that Shuriken were all killing implements, far from it. Though many were given an edge nearly as sharp as a sword, the vast majority remained relatively blunt affairs – due as I have already stated, to quick manufacture. Unlike the movie actors of today would have us believe, Shuriken or Shaken were rarely death stars.

There is a theory which is referred to as the "deity cause", which plants the origins of Shuriken and Shaken firmly in the age of the Gods – the mythical forerunner of Shuriken being the Vajra thunderbolt of Aizen Myo-o, and the Shaken being a representation of the Horin wheel of Kharmic order.

Still more intriguing is the Uchi-ne, the hand arrow – possibly developed on the battlefield. A Samurai, with swords gone, bow torn asunder, could ably defend himself with a broken arrowshaft. In time the Uchi-ne came to resemble a miniature spear with a tassle to aid its flight. These weapons which we see today, may be of seventeenth century manufacture and development.

This is so important a factor, that I pause to include this, in the spirit of "watch out, you could get caught", as Sasano Tomisaburo delighted in repeating. I have seen in the collection of a good friend of mine, a scholar of no small standing, a number of composite weapons of fierce appearance. When I questioned him about them, he let a wry smile cross his lips, then put me out of my misery – A great many such pieces were made for gentlemen Samurai and rich merchants who aspired to be Samurai. They were manufactured in the late seventeenth, early eighteenth century, when battles were a thing of the past. A whole industry grew up to slake their thirst for warrior equipment. These are a few examples, so as my friend says, "watch out, you could get caught".

*Umabari – Higo Province.*

*Mitokoromono (things for three places). Shishi (liondog) after the style of Goto Tokujo, Momoyama Period.*

**Example One** – A suit of armour, made two hundred years ago containing literally everything in the development of armour. The whole piece looks magnificent, but I regret, were it worn in earnest, its shortcomings would immediately become apparent. Most notable is the inclusion of a socket at the back for the Sashimono battle standard. Unfortunately, coupled with a massive Kabuto helmet in the style of an earlier period, the poor victim would be forced to look permanently at his feet, his head crushed forward by the Sashimono butting against the Kabuto.

**Example Two** – A Keibo, studded warclub of hexagonal form, with a musket rest at one end of the handle. Contained within the Keibo was a straight musori single-edged sword. When I made a comment, my host merely said, "funny business".

**Example Three** – A sixteen-fold iron Tessen or Tetsu Ogi, iron fan, containing a small shuriken-type Kogatana and a Jitte sword breaker. The provenance of these items must, as stated before, be assigned to early Tokugawa-1750's. Their absolute effectiveness being in question. In fact, in Arai Hakuseki's "Honchō Gunkiko", a treatise of Samurai equipment, produced in the seventeenth century, these anomalies abound.

It is said that they were an invention to slake the thirst of a generation who never knew the cold excitement of the dawning of a battle. Be careful then, that you are not "taken in" with Shuriken and Shaken. Many have fanciful names, appended by modern exponents. I leave this to your judgement, whether you believe it or not.

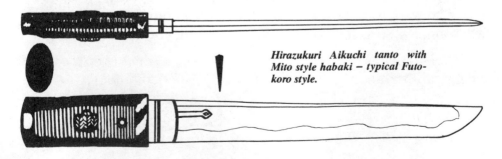

*Hirazukuri Aikuchi tanto with Mito style habaki – typical Futo-koro style.*

CHAPTER SIX

# Preparation for Practice

*"Though I shall look for a thousand ages
I shall not be wearied"*

MIROKU

*Kasa Kanamura 725 A.D.*

Preparation cannot be stressed too much. There are those who advocate "getting straight into" the throwing. But here I counsel caution, action in haste is always regretted. As a building cannot be magnificent if its foundation is lacking, thus it is with practice. Whether your requirement is classical, modern cognate or modern synthetic, preparation is the key word.

**Clothing**

If the student is a practising member of a classical martial Ryu, utilising Shuriken Jutsu as an essential part of their curriculum, the rules for correct dress are those set by the individual Ryu. In general terms, this comprises, a jacket, similar to that worn in modern Kendo, ie with a single tie at chest height; the colour being white, black or dark blue. The ensemble being completed by the pleated culotte known as hakama, again in either white, black or blue. The combination of colours, being the choice of the Ryu. Feet are usually bare, though some Ryu favour the split-toed jika tabi boot for outdoor wear.

Adherents of the modern cognate tend to favour their style's "Gi". Thus a judo player, would wear judo gi, a karate player, karate gi, and so on. They may supplement this with hakama, but this is a personal preference.

Practictioners of the modern expression of the feudal ninja, tend to favour the all black rig and mask, as popularized in many motion pictures.

Followers of the modern synthetic may choose to wear any of the former, though many adopt the track suit and sneaker or even the "vet" combat fatigues and high boots. I make no further comment as to these latter practices, as they are in the process of developing an identity. Time, as always, will be the test of these synthetic forms.

Generally speaking, clothing should be loose enough to avoid con-

*Roku Hoshi-Sukashi.*

*Sankaku-Sukashi.*

stricting natural body movement, but free of loose ties or hanging zippers or anything which could conceivably snag or catch a Shuriken or Shaken in the act of throwing.

### Choosing the right Shuriken or Shaken
This of course depends a great deal upon which form of Shuriken Jutsu is practiced. In general, the best advice is to go for quality. Shaken stars, with an enamelled "portrait of an actor" or various oriental-looking signs are interesting as memorabilia, but have little to do with serious practice. I would strongly advise those of the best quality steel manufacture.

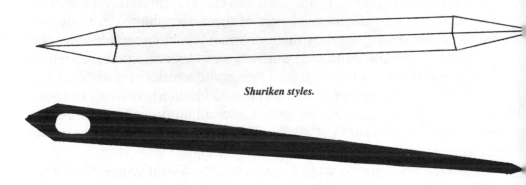
*Shuriken styles.*

### Care of Shuriken and Shaken
It depends very much what finish is on the item, as to what is the best advice to give. Stainless steel needs little cleaning, save an occasional degrease and polish. If, however, you ascribe to the purist view; black steel is the norm. In cheaper varieties, this is achieved through spray-painting which will quickly chip to reveal bright metal underneath. The best form is that which has a "black rust" coating. This can either, as with the antique forms, be achieved over the years. Or by use of various chemical pickling processes (these are often best known to the manufacturers, but the aspiring homeworker may, with recourse, to a chemist, achieve a similar result).

If however, one is fortunate enough to possess a black rust set, the care is quite simple:– After use, cleaning with a light cloth to remove any traces of the oils and acids which naturally accumulate in the sweat and cause red rust. Red rust is the most dangerous, for it is the most corrosive form, eating into the very metal itself. Many afficionados choose to oil their Shuriken and Shaken with a little choji abura, this is a delicately perfumed clove oil, which is essential in the Teire (maintenance) of Japanese swords.

A word of warning – often, particularly in the west, I have seen items literally dripping with oil and grease. My advice – whether it be

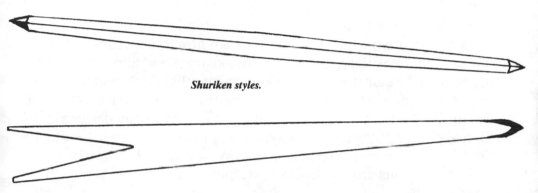

*Shuriken styles.*

Shuriken or sword – a little oil judiciously applied and rubbed off with a soft cloth is best. If the metal is left covered with this excessive oil or grease, the following will occur. The oil will darken and form a laquer-like coating which is very difficult to remove, underneath will form a most virulent "oil rust".

If you acquire a set of old Shuriken or Shaken, my advice is clear – do nothing in haste. The indiscriminate use of wire brushes, steel wool or heavy metal polishes will certainly clean the item, but at the cost of its original patina, which in the antique form, was a product of years of patient care.

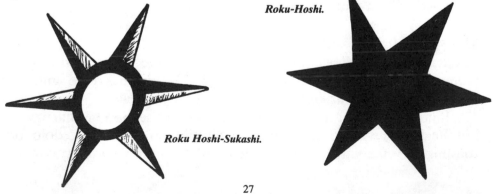

*Roku-Hoshi.*

*Roku Hoshi-Sukashi.*

*Sanaku-Ho.*

*Shikaku-Ho.*

*Manji.*

One "trick of the trade", which I can reveal without undue worry, is to hang the Shuriken or Shaken by hemp or paper twine (anything which does not have an acid/dye base). Traditionally, this practice has been known as "Nokishita" – under the eaves, so that it is protected from the direct onslaught of wind and rain. In Japan, the traditional house design accommodates this without undue difficulty. Those of my readers who are outside Japan must, I am afraid, experiment. But do not use your treasures; instead, experiment on an old Shuriken which is not needed. Usually fourteen to eighteen months is sufficient to coat the item with a silky black patina, which is most pleasing to the eye and to the touch. At this point, I must warn you to thoroughly de-grease the item and wear cotton gloves so you do not leave acid sweat – often a Shuriken or Shaken has been hung for over a year, on taking it down, its owner has, to his dismay, discovered a perfect red rust facsimilie of his finger-prints.

### Storage of Shuriken

When not in use, Shuriken or Shaken should not be left rubbing against each other in the manner of small change. Some purists keep their Shuriken and Shaken in fitted pawlonia – wood boxes made specifically for the purpose. They go so far as to separate them with sheets of Hosho paper, I realise that even in Japan, this may be difficult to obtain, a well-known brand of tissue paper is perfectly acceptable. On no account use cotton wool or leave on a wooden surface. Whilst it may look "Japanesey" to display your treasures in this way; the close contact with the wood will encourage red rust.

### Transport of Shuriken

There are certain areas, particularly in the United States, where the possession of Shuriken or Shaken is illegal. My advice is quite simple – do not transport Shuriken or Shaken into these areas.

If however, you are in an area where the possession of Shuriken is not illegal and wish to transport your Shuriken from home to dojo for training and vice-versa, I urge you to adopt the following safety code:–

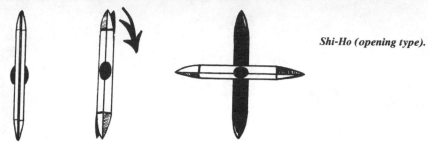

**1** Always carry Shuriken in a box or case, which has a definite closure.

**2** Shuriken must not be carried loosely in a pocket or in a belt-holster. This is in contravention of most ordinances concerning concealment of weapons.

**3** Carry the box or case wrapped up in a towel, which is in turn wrapped up in your training kit, at the bottom of your bag.

**4** Understand the point of view of the law-enforcement agencies. Their's is a difficult task and an understanding of their problem in differentiating between a weapon concealed and a training implement safely stowed, will go a long way to avoid a situation arising.

**5** Always carry your dojo and association membership documents.

**6** Only transport to and from dojo for the purpose of training.

**7** Ideally, if conditions permit, leave your equipment at the dojo.

**8** Do not "boast" about your skill with Shuriken or offer to demonstrate.

**9** The concept which some so called, instructors, peddle of "Shuriken street defence" is both stupid and dangerous. Do not attend such classes, they are probably illegal.

**10** Avoid storing your training kit in the back of your car for long periods of time – Do not assist the criminal in unwittingly obtaining weapons.

With a little forethought and common-sense your study of Shuriken Jutsu, no matter what persuasion, will be a fruitful and stimulating experience.

# CHAPTER SEVEN

# Stances and Grips

SHAKA

*"On the snow*
*Alighting gently*
*The nightingale"*

*Kawabata Bosha – Meiji Period*

Regardless of persuasion, whether the practictioner be of classical, modern cognate, or modern synthetic, the basic throwing stance varies very slightly. It can best be described as a variation of the Karate back stance, known as, Kokutsu Dachi, with the weight centring on the back foot. In the action of throwing, the posture somewhat resembles the Karate front stance of Zenkutsu Dachi, with a percentage of the weight, centred on the front foot. There is a great deal of misunderstanding and misinformation; most of it well-meant.

Within the classical Ryu, the naming of particular stances are bound up with the deep martial philosophy of that particular Ryu. Unless you are a bona fide member of a classical Ryu, it makes little sense to copy the classical stances, as, depending upon level, they may be inherently designed to disguise their actual effective use. It has been popularly documented that this is indeed the case of Tenshin Shoden Katori Shinto Ryu, whose entire Omote No Tachi Kata has a facet known as Kuzushi. This analysis of the form has great depth – for example, in the pairs bokken (wooden training swords) practice; bokken are struck noisily and certain movements deliberately omitted. In battle reality, the sword would glide noiselessly. This deliberate deceit is yet another example of the high standards required within the Ryu – a student is tested constantly.

For those of you who are interested in this subject from an academic standpoint, I can heartily recommend a series of three books, which are available outside of Japan. They were produced by Minato research and publishing company, their title is:–

*"The Deity and the Sword", Volumes I-III*
Written by, master teacher Otake Risuke
Translated by, Donn F. Draeger and Shinozuka Terue
and Nunokawa Kyoichiro
Photographed by, Hirata Sadao
Edited by Sugawara Tetsutaka and Ando Takako

I make no apologies for this totally unsolicited testimonial, for the quality of the series is of the highest order.

Stances in the modern cognate forms are governed by the style origin. Thus, an Aikido practictioner may tend to favour Sankakutai, the basis of Aikido. Likewise, a Karate Ka would choose a Kokutsu Dachi of high or low form, depending upon school origin. A Shotokan influenced Karata Ka, utilising modern cognate Shuriken Jutsu, will do so from a low Kokutsu Dachi, involving a great deal of major muscle groups. On the other hand, a Shukokai – influenced Karate Ka, will adopt a high, swift Kokutsu Dachi in his or her practice of modern cognate Shuriken Jutsu.

The situation regarding the modern synthetic is one of flux. As previously stated, these forms are in a process of evolution, but in general, the favoured stance-names are, "straddle", "T. Form" and "front fire".

If you are without the aid of an instructor in any of the aforementioned disciplines and wish to study Shuriken Jutsu, I can offer some basic advice which I hope will be of assistance. It may, to the purist, seem a "rough and ready", method, but it is a safe and sure method.

## Stance in practice

You must familiarise yourself with the mechanics of stance before going on to the more interesting work. This is preparation and no viable system can be without it.

The most basic positions, Dachi in Japanese, that will enable you to progress by yourself are:–

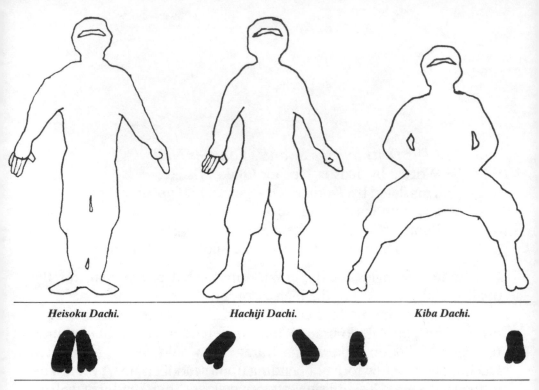

*Heisoku Dachi.*       *Hachiji Dachi.*       *Kiba Dachi.*

<u>HEISOKU DACHI</u> – a natural, relaxed stance with feet close together.

<u>HACHIJI DACHI</u> – legs apart, feet shoulder width, with toes pointing outward at 45°, body upright and relaxed.

<u>KIBA DACHI</u> – keep your back straight, with your feet two shoulder-widths apart, with the toes facing forward. Bend the knees evenly until they are just forward of your toes.

<u>KOKUTSU DACHI</u> – bend rear leg and tense it outward; support 70% of your weight. Bend your front leg slightly at the knee. Point your toes forward.

<u>ZENKUTSU DACHI</u> – bend your front leg forward, so that 60% of your weight is on it. Your rear leg should be thrust back at an angle of 30°, approximately two shoulder-widths apart.

The following drill will give you a good grounding in the body work, that is so essential. You may place your hands on your hips, which will give you an idea of how important the rotation of the hips is – study this point well.

*Kokutsu Dachi.*　　　　　　　*Zenkutsu Dachi.*

Stand in Hachiji Dachi, move forward with your left leg into Zenkutsu Dachi. Hold for a count of two, transfer weight onto back foot to produce Kokutsu Dachi – hold for a count of two. Turn to oblique right in Kiba Dachi, hold for a count of two. Move back into Zenkutsu Dachi; hold for a count of two. Then step back into Hachiji Dachi. Repeat the whole exercise on the right foot.

The more you practise this drill, the greater will be your understanding of hip rotation and hip movement. As you become more accustomed to this movement, you may vary the speed, similar to the combat practice speed which will be dealt with in another book.

*Method of holding Shaken.*

# Grip

It is reasonable to state that the grip of the Shuriken or Shaken, varies with the form of practice. For example, within the classical Ryu, there are numerous types of grip, some of which, as described with reference to the stances; are deliberately misleading. Grips such as In No Ken, Yo No Ken, Ryu No Ken, Ishi No Ken, Kusanagi No Ken, being the most popular across the Ryu.

In the modern cognate forms, the overall term of "Te" seems to be the most popular form. As will be seen by the illustrations. Te is appended with Mae – forward, Ushiro – rearward, Yoko – sideward.

The modern synthetic tend to be more pragmatic in their terminology – with such grips as, point forward, point backward, and star sideways.

It is well to remember that the grip will vary with the style of Shuriken or Shaken used. Again I speak to those of you without the aid of an instructor. The basic advice I offer is this:–

> Point away from your hand for medium distance.
> Point towards your hand for long distance.
> If you are using Shaken, keep the star canted
> slightly backwards, so that the tines do not
> snag on the release.

You must research this thoroughly. Also, I urge you to experiment, as the individual Shuriken may, in flight, reveal an anomaly of weight distribution, not readily noticeable in normal study. You must strive to make the Shuriken or Shaken as much a part of you as possible. A grip should feel comfortable. If not, you must discover the reason – Are you gripping too tightly with the thumb or forefinger? Too tight a grip is a common fault.

As a matter of interest, I include the following:–

It is useful to note that not all projectiles were thrown by hand. The split-toed 'tabi' socks afforded a skilful Samurai yet another method of distracting an enemy. A small Shuriken or Shaken, or even a small stone; could be held between the big toe and its neighbour. By flicking the foot forward, much in the manner of Mae Geri Keage, the projectile would head for its target.

Yet another method, was to grasp lightly, between the lips, what can best be likened to an iron tooth-pick, sharpened at both ends. This noisome weapon was held horizontally between the lips. By taking a deep breath through the nose and breathing out through the mouth, with the sound – PAH! the item could be expelled with great force. As much as two and one half meters with accuracy, could be covered. Certainly not with a deadly accuracy, but enough to allow a sword to be drawn or a hasty retreat to be effected.

Carefully observe the illustrations showing grips. It will be necessary for you to practice for many hours in order to get the "feel" of the Shuriken or Shaken. It is a good analogy to imagine that the Shuriken is made from an extremely fragile material. In time this will give the "feel", which enables a smooth, accurate throw. You must study this point deeply.

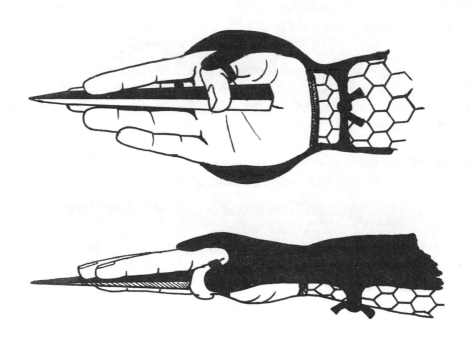

*Method of holding Shuriken with point outwards – note Hyogo Gusari chain mail, typical of Senyoku Jidai.*

# CHAPTER EIGHT

# Throwing and Targets

*"The moon in the cold stream. Like a mirror."*

*Miyamoto Musashi. 1645.*

HAGUNSHO

At this point I must assume that you have thoroughly practised the basics of stance and grips and are relaxed with them. It is a pointless exercise to rush all the major points or to try to compress the experience. No matter how swiftly your intellect feels it understands – your body will not. There is a marked difference between body and intellect; you must continue in the basics, even when quite advanced; so that you can truly express the Samurai maxim of,

*"To know and act are as one."*

For the purpose of this volume I will deal with only three actual throwing techniques. I feel it counter-productive to have you practice any more at this stage. All the following instructions are for the right hand, although the left hand can also be used.

**Throwing method number one:**
This can best be described as the basic form.

**1** The throwing hand is raised above the shoulder, approximately level with the top of the head.
**2** From this back position, the hand is propelled palm-forwards until the natural arm extension is achieved. This will propel the Shuriken forward.
**3** Increase of power is achieved by the counter-action of the other hand, which resembles the hikite of Karate.

This throw has many names, the most popular being, O-Uchi, Seijo Uchi, Yatto Uchi, and hon Uchi. I prefer hon Uchi – refer to Kata.

**Throwing method number two:**
Adherents of modern Iai Do will recognise the fundamental similarity between the action and that of the Nukitsuke of basic Iai-Do form –

*Demonstration of Hon Uchi throw without body shift.*

refer to illustration. Basically the action, as in all of the three throws explained in this volume, is very simple.

**1** Start off with the Shuriken held in the right hand with the palm turned downwards.

**2** The thumb edge is on a line with the top of the Obi (belt) on left side. But in classical form, above and away from the swords.

**3** The throwing action is quite straight forward. The arm rises up and out across the body to shoulder height. The palm remains downwards at all times.

It is known as Nukitsuke, Kiri No Ken, Gyaku Uchi: I prefer Nukitsuke.

*Breakdown of Nukitsuke throw.*

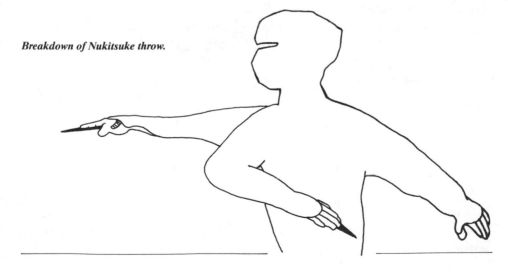

## Throwing method number three:

This third method is really a variation of method number two, with the exception that the arm does not rise to the loosing position.

**1** The Shuriken is again held palm down in the right hand.
**2** The hand is held at shoulder height, the thumb edge close to the left shoulder.
**3** The throwing action is accomplished by moving the right hand horizontally forward, palm facing down still – to the release point.

It is known as Ichimonji or Ichi, because the horizontal form of the throw resembles the single horizontal brush stroke that represents the numeral one (ichi) in Japanese. It is also known as, Yoko Uchi: I prefer Ichimonji.

*Breakdown of Ichimonji throw.*

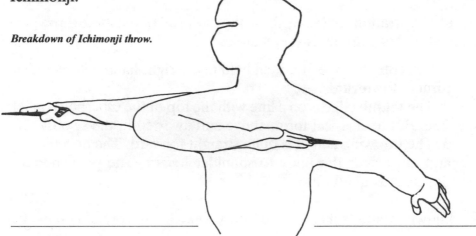

## Throwing Shaken

The three Shuriken throwing methods are also applicable to Shaken. The only point I will make is to be aware of the "bias" on the throw which occurs with Shaken. You must experience this to understand what I mean. The remedy is in the angle of the hand to wrist – only practice will reveal this, so I advise the keeping of notes.

## Special strength training

> WARNING – it is advisable to consult your
> doctor before attempting this practice.

Using a Tessen, iron fan or bar of about six and a half pounds weight, the action of the drill is to go through the elements of each throwing action, without, of course, throwing the Tessen. Be careful, the weight is decep-

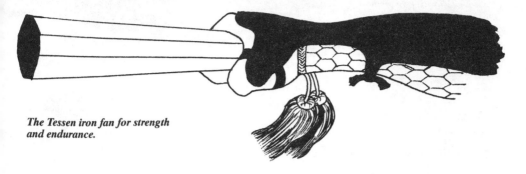

*The Tessen iron fan for strength and endurance.*

tive. Care must be taken to avoid undue strain at wrist, elbow and shoulder.

Start slowly, five repetitions to begin with; then finish with ten complete throws with Shuriken. The effect is to strengthen the elements of the throw; as a side-effect, endurance and recovery time also improves.

## Distances

I am afraid there are many "fancy" methods for distancing; specifying to the very centimetre, the distance from the target. I am sorry to disappoint any of you who expect this from me. My advice with distance is simple; keep close when practicing basic multiples. As your skill increases, you may move further away each time. You must experience this for yourself in terms of paces. Again keep notes – you must formulate and evaluate your own progress. There is no set distance because there are so many variables eg. length of arm, elbow to wrist measurement, centre of gravity etc.

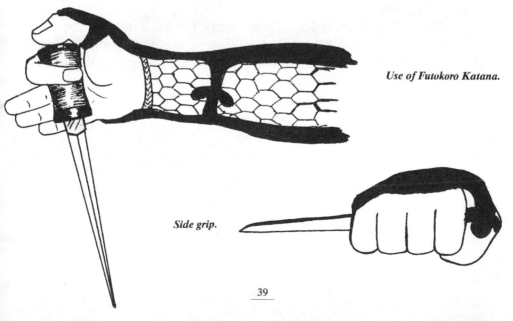

*Use of Futokoro Katana.*

*Side grip.*

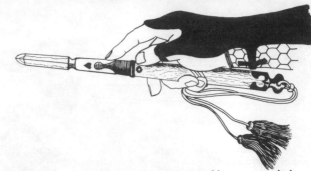

*Uchi-Ne (Koyari) grip method for throwing.*

## Aiming

This is dealt with in the next chapter, I make no mention of it now as it is inextricably bound up with the gaze in the Kata form.

## The point of loosing

There are those who would make a grand affair of the moment of loosing; likening it to the great liberation of Zen archery. I cannot comment on this with reference to Shuriken Jutsu, except to say, "don't think, just throw".

## Targets

Safety is the byeword. If in doubt, do not throw. It is your responsibility to check before throwing. Don't expect others to know what you are thinking, always announce your intention to throw in a clear voice. If you are awaiting your turn to throw, keep well behind the performer; do not make any unnecessary noise or movements which may distract a performer. In many classical dojo, those waiting to perform are required to sit in Seiza in a line slightly behind and at 90° to the performer. It is no coincidence that in the classical dojo, accidents are so rare as to be hardly spoken of.

Targets for Shuriken Jutsu take many forms. There is the compressed fibre "dartboard" type, which utilises the soft-edged safety style indoor type. In Japan, the use of the straw makiwara, similar to that used in Kyudo is much used. Certain styles can be hung in the trees and

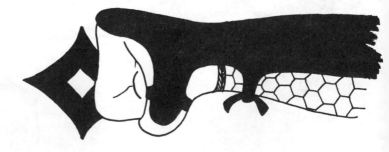

*Uraken type of grip – flicking action.*

*Alternative grip.*

swung to afford a moving target. Many practitioners use a sawn tree trunk. I add, as a matter of course, that the habitual use of a particular living tree trunk as a target is both unnecessary and wasteful, as in time, the tree will become diseased, leaving your garden the poorer.

I have, in my own garden, a sawn tree trunk of the Yanagi willow. This is sunk into the ground, much in the same was as a traditional Karate makiwara. For those of you unfamiliar with this method, I have had my illustrator include a drawing which may be of some use.

Of course, I realise that those of you in zones where the willow will not grow, may be at a disadvantage. I cite the example of an acquaintance of mine who resides in the United States of America, who suggests the use of telegraph poles or railroad ties (sleepers). This method is perfectly acceptable, but be warned against an over-tarred pole as it may cause binding of the Shuriken.

Those of you who do not wish to use a heavy tree trunk form, for reasons of space and economy may use the following:– worn-out tatami, solid wood doors, builder's fibre boards, in any material which will accept Shuriken. If indoors, keep fairly close to a wall with a firm base – targets must not rock. If outdoors, a curtain behind the target will catch any strays.

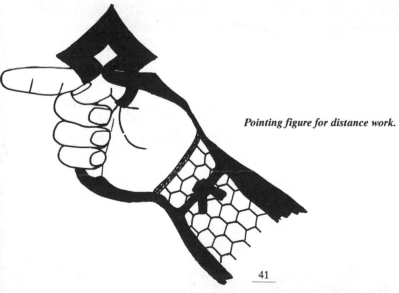
*Pointing figure for distance work.*

*Grip for heavyweight Shuriken.*

Many people prefer to use Shaken with a hole punched in them; not for the old myth of it being the method of carrying them on a stick in the belt (they were more probably strung like beads on a cord similar to Sageo) or for their handling qualities, but because if thrown too hard at a target, they can be removed quite simply by the use of a tool which has been developed for the purpose. It is simply a hardwood dowel of a diameter slightly less than the size of the hole in the Shaken. At one end of the tool is a metal cap; by passing the dowling through the Shaken and by exerting a steady pressure, in a levering action, the Shaken comes safely and cleanly away – thus avoiding the use of pliers etc, which mark and distort the edges.

Often with serious practice, the edge of Shuriken or Shaken may become dulled or chipped. This is a typical effect in a tempered edge. Ideally they should be sent to an artisan who is used to the precise angle-honing which is necessary for restoring the edge and point. Purists tend to favour Amakusa stone, this comes in several grades and is readily available for household sharpening of kitchen tools. Those of you living outside of Japan might try to obtain Amakusa in one of the stores catering for Japanese communities. Failing that, the finest grade carbo-rundum block, such as washita can be substituted.

The action with Amakusa is to use a liberal amount of water, in the Japanese method, the tool is drawn at the precise angle towards the body. The action with carborundum block is with a liberal amount of fine oil, the tool being pushed away from the body at the required angle.

A Shaken star which has bent out-of-true, can be gently coaxed back; unless you know the precise forging details, avoid re-heating. Place the star flat on a block of soft wood, larger than the star. A similar block placed on top and struck with either a flat mallett or hammer (flat-face) in a circular action until the flatness is achieved.

I cannot conclude this chapter, without reiterating the warning with which I started it. When throwing – If in doubt, don't throw.

*Safety first at all times.*

# CHAPTER NINE

# Mind Control and Kata Form

*"Unawares, all grew old:*
*Even, the mountain's cypresses, standing like spears"*

*Komo Taruhito*

DAI NICHI

Now that you have a rough basis of a workable Shuriken Jutsu form; I digress slightly from the physical expression to delve a little into the mental processes which are common to both classical, modern cognate and perhaps, depending upon the instructor, modern synthetic.

**The gaze:**

Unlike rifle or pistol shooting, where targeting is the epitome, in Shuriken Jutsu, specific targeting is of little or no use. One should instead, aim for an all-encompassing gaze. Your facial muscles should remain relaxed; obsessive staring or a "macho" tough look fades into weakness when confronted with the real gaze of spiritual strength which emanates from inside, not outside. Study this well. You may try an experiment. Face the target and "psyche yourself up" by grimacing and making what I can best describe as a violent mind – similar to 'psyching up' in American football – Now throw ten Shuriken and observe the results and how you feel. Now give yourself ten to twenty minutes' relaxation, perhaps by listening to some soft music. Concentrate on breathing softly and slowly. In this frame of mind, face the target again. Throw ten Shuriken, not caring about whether they hit. Again observe your results and how you feel. I advise keeping a notebook. Jot down these tests, perhaps every month as you progress. At first your "psyched-up" angry mind will get more satisfaction and accuracy. But as you progress, in your mental conditioning, the reverse will come about, until you will be able to throw fast and accurately with a calm mind. You must persevere in order to obtain these results; just trying once and dismissing it as rubbish, is not the spirit of the true martial artist. Often, things are deeper than at first observed.

The gaze which I am talking about is known as Enzan No Metsuke.

It can be likened to looking at something in the far distance. As your gaze does not centre upon anything within your near sphere, you are not disturbed, your spirit is calm – but you are totally aware of everything that is occurring, but you are not affected by it. It is useful to acquire this skill. Outside of the dojo, in business, the use of Enzan No Metsuke is most beneficial, as when dealing with complex negotiations you will not become easily side-tracked. I urge you to try it.

**Breathing:**
There is much lip-service paid to breathing techniques, but very few martial artists practice more than the basics – regarding them as "just for the beginners". This is absolutely <u>not</u> the case. The more advanced a person becomes, the greater his need to study and practice breathing

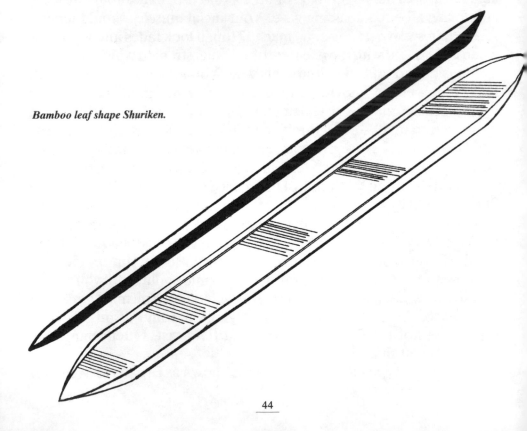

*Bamboo leaf shape Shuriken.*

techniques. It must become, as it were, his core of study. I include a few practical methods which will be of use; no matter what the system of practice.

WARNING – Consult your physician before attempting this practice

**Method Number 1**
This comes from the Soto School of Zen, which, because of its stress on the importance of non-attachment, became known as the "Samurai's Zen".

Sit in seiza, or stand in Hachiji-Dachi. Relax your shoulders, let your hands hang loosely at your sides. Take a soft breath through your nose. It should make the sound, SOO! The duration of your in breath

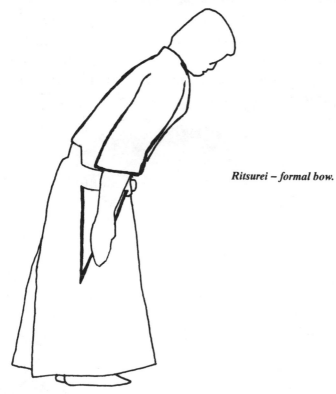

*Ritsurei – formal bow.*

should ideally be thirty seconds. I realise that this may be a little uncomfortable at first. Start with 10, then 15 seconds until you reach the desired time. With the lungs full of air, hold the breath for a count of between 5 and 10 seconds. Then breathe out through the slightly open mouth. Keep your facial muscles relaxed, do not force the breath out; instead, let it flow out gently with a deep sonorous Ra! This again should take thirty seconds. As before, start off gently and build up to the optimum time. When you come to the end of your single exhalation, pause, put strength into your lower abdomen and force out the residual volume of air. Pause for a count of ten, then breathe in and repeat the sequence.

> *WARNING – The times contained here are the ideal figures. Be careful, you may feel light headed at first. This is quite common; stop, rest, then begin again.*

If you experience tightness in the chest, or shoulders, do not worry, this is because you are "chest breathing", you must use the intercostal muscles and diaphragm. By swelling your stomach as you inhale, you will accomplish this without undue strain.

**Method Number 2**
This comes from the Misogi ritual of purification in Shinto.
Stand in Hachiji Dachi, arms hanging at your sides. Spread your fingers wide apart. As you inhale through your nose, bring your arms up, using the shoulders as a fulcrum, much as a lotus flower opens and closes. When you have inhaled, your arms should be at 45° to your head. Pause for a count of ten, then start to exhale with the sound, YAH! As you do so; with a feeling of pushing down, return your arms, fingers spread, to the original start position. Keep your gaze, Enzan No Metsuke. Repeat whole sequence.

**Method Number 3**

This is a Sogo-synthetic method, based upon a Misogi practice. It is known as Aun no kokyu.

Crouch down, trying to make yourself as compact as possible. Breathe in to the sound, "Ah", through both nose and mouth. When inhalation is complete, pause for a count of ten, then reach upward from the crouching position; breathing out through the mouth with the sound, "Un". As you do so, reach up with your outstretched fingers, as if trying to touch the sky, hold for a count of ten, then repeat. Be very careful, this is a very advanced breathing method, work up to it slowly.

**Method Number 4**

This is Soto-Zen style meditation, Za-Zen.

Sit comfortably in full, half-lotus or seiza, or in a chair if these positions are not possible. Direct your gaze approximately $2\frac{1}{2}$ metres in front of you. Purists place their hands in the manner of the Buddha, but you may rest your hands gently on your thighs. Close your eyes slightly. Begin a regular pattern of inhalation and exhalation – in through the nose and out through the mouth. On the exhalation, count one and so on upon each exhalation until you reach one hundred. If you forget, or think that you have made a mistake, start the count from one. Try it, it is more difficult than you imagine.

These breathing methods have a beneficial, calming effect on the mind and body and serve to calm one down after a hectic day. Thus one enters one's practice of Shuriken Jutsu, rested and refreshed.

Remember, if you feel lightheaded, or in any discomfort, stop. The purpose of the exercises are to develop, not mortify.

As you will see, there are many and varied breathing/meditative methods. I will explain more of this at a later date. Suffice it to say, within the context of this volume, without some special breathing/meditative regime, the true in-depth study of Shuriken Jutsu, whatever its provenance, cannot progress.

## Concepts within practice

*Onore O Sameru*, the highest pinnacle of Budo, to control the self, even in the face of the inevitable nemesis. This is best exemplified by the example of Grand Admiral Yamamoto Isoroku, who foresaw the cataclysm, but bravely continued, in the spirit of Yamato Damashii; I echo his momentous words –

> *"I am the sword of my Emperor, I will not be sheathed until I die"*

## Satsujin No Ken

Satsujin No Ken can be seen as a mental process, whereby one becomes a weapon solely for the purpose of taking human life; literally annihilating anything in the way of one's goals.

## Katsujin No Ken

Katsujin No Ken is the mental process whereby one's weapon becomes life-giving. It is a difficult concept to understand, for it demands total destruction of the ego.

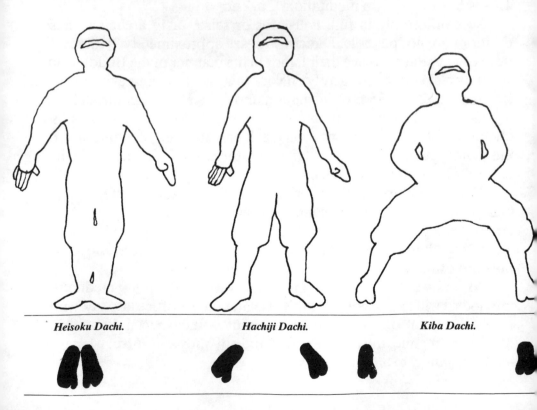

*Heisoku Dachi.*     *Hachiji Dachi.*     *Kiba Dachi.*

## The Kata

By the very nature of Shuriken Jutsu practice, its very soul lies in Kata, regardless of the persuasion. In the classical Ryu, Kata are as numerous as the Ryu themselves, for with few exceptions, training in Ko-Ryu (martial Ryu) consists of progressively more difficult Kata, which contain as it were, the heart of the Ryu's battlefield-proven methods. At the same time maintaining the spiritual code which is the very cement of the Ryu's moral ethic. It would be misleading in the extreme to document the Kata, for, as stated elsewhere; there is no representative form. Many Ryu have Kata names which are the same in sound, but totally different in both physical and spiritual content.

Names such as:–

| | |
|---|---|
| Ten No Kata | *Form of Heaven* |
| Taikyoku No Kata | *Form of Essentials* |
| Manji No Kata | *Form of Eternal Happiness* |
| Chi No Kata | *Form of Earth* |
| Jin No Kata | *Form of Man* |

These seem to be the most popular names and they occur in Ryu all over Japan.

*Kokutsu Dachi.*　　　　　*Zenkutsu Dachi.*

The Kata in the modern cognate form however, seem to be a some-what haphazard affair, though I will cite a few more workable examples which will be of constructive use. I have taken the initiative in this area and labelled them as, Taikyoku, meaning essential. I have devised ten which increase in complexity. I shall include two in this book, they are; Taikyoku No Kata Ipponme and Taikyoku No Kata Nihonme.

Whether the modern synthetic drills can be called Kata I am doubt-ful. By their origin, they are subject to some dispute, therefore I will in-clude them in chapter ten, where I offer them as an interesting diver-tissement. It is my conscious choice to keep them separate, for reasons which will become readily apparent.

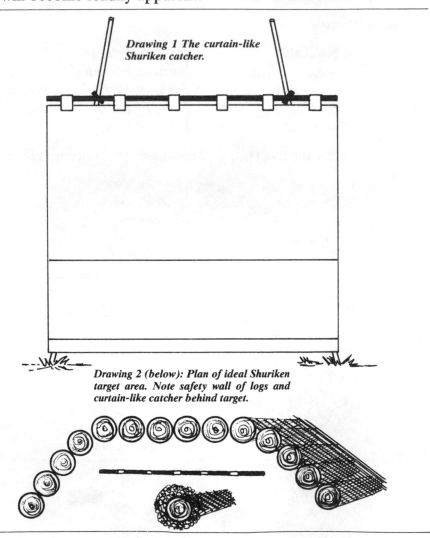

*Drawing 1 The curtain-like Shuriken catcher.*

*Drawing 2 (below): Plan of ideal Shuriken target area. Note safety wall of logs and curtain-like catcher behind target.*

## Basic Kata Number One

*Taikyoku No Kata Ipponme:* Left or right.

*Phase One:*      Approach target; Shuriken in opposite hand to throwing hand.

*Phase Two:*      Stand in Heisoku Dachi. Execute Ritsurei (formal bow) to target (Tip:– do not take your eyes off the target).

*Phase Three:*      Step into Hachiji Dachi, moving opposite foot to throwing hand.

*Phase Four:*      With a deliberate action, take a Shuriken and place it in your throwing hand, in the favoured grip for the particular style of Shuriken.

*Phase Five:*      Step forward into Kokutsu Dachi. Make preparation for throw. I remember the advice of my old Gekken teacher – always step forward never backwards, and in a hundred battles, you will never suffer defeat. Really he meant to instil the concept of *"continue, no matter what the obstacles"*. Where there is a choice, never pause; step boldly forward. This is a very deep concept, I urge you to study it deeply, for it also pervades life.

*Phase Six:*      Maintain Enzan No Metsuke but be aware of target; Hand and Shuriken. Calm yourself (relax facial muscles, drop shoulders, envigorate lower abdomen).

*Phase Seven:*      Adopt 'L' form ready position, still in Kokutsu Dachi.

*Phase Eight:*      Execute Hon Uchi (Yattoh) strike. As you do so, thrust your weight forward into Zenkutsu Dachi. Tense your lower abdomen and as the Shuriken leaves your hand, you must execute Kiai – the "spirit shout" which is the result of the unification of the physical and mental. The sound should well up from inside yourself as the sound, "Yah!", "Ei" or "Toh!" It should be a deep sonorous sound. Don't scowl. You will only discover this depth of power from the rigorous breathing practice, described earlier. I will say no more, it is up to you to liberate this ability.

| | |
|---|---|
| *Phase Nine:* | Maintain the follow-through gaze, still Enzan No Metsuke. Inhale quietly. Throwing hand still at Hon Uchi (Yattoh), end position. The feeling which pervades your whole existence is that of Zanshin – total alertness. There should be no Suki – gap in concentration, because, as the old Samurai maxim goes – if there is one inch of gap, one inch of water may fill it. Thus are mountains brought to dust. |
| *Phase Ten:* | Move hips back into Kokutsu Dachi, move hands down from Hon Uchi (Yattoh) position. Take another Shuriken and repeat phases 6 through 9, until all Shuriken are thrown. |
| *Phase Eleven:* | Until last Shuriken is thrown, maintain Enzan No Metsuke with Zanshin mood. Step back from Zenkutsu into Hachiji Dachi (symbolically to step back at this point is to offer no agressive intent). |
| *Phase Twelve:* | From Hachiji Dachi, adopt Heisoku Dachi. Hands lightly against your sides, execute Ritsurei, to both target and Shuriken. You may now retrieve your Shuriken. The Kata is complete. |

(No matter how many repetitions of Kata are performed, do not dismiss Ritsurei as a waste of time. It is designed to calm the mind for the task in hand.)

*Modern ratchet double-face swing target.*

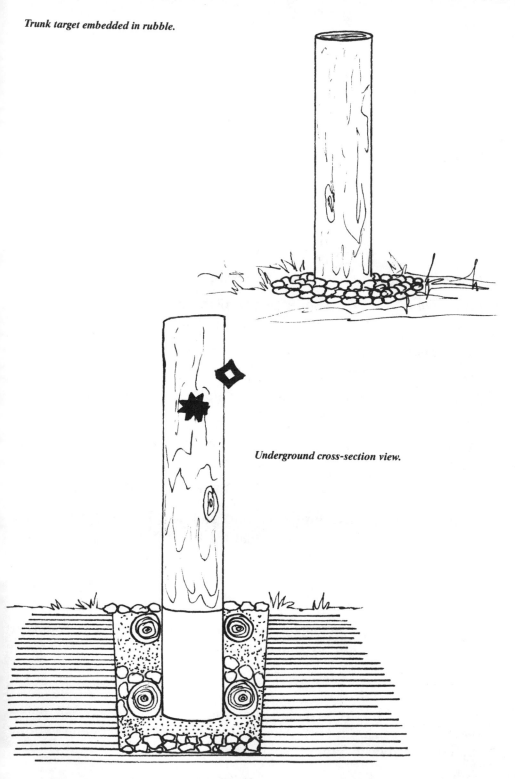

*Trunk target embedded in rubble.*

*Underground cross-section view.*

53

## Basic Kata Number Two
### Taikyoku No Kata Nihonme

*Phase One:*      Approach target, Shuriken in opposite hand to throwing hand.

*Phase Two:*      Stand in Heisoku Dachi. Perform Ritsurei.

*Phase Three:*      Step into Hachiji Dachi.

*Phase Four:*      Take a Shuriken and place it in your throwing hand in a grip suitable for a Nukitsuke cross-draw throw.

*Phase Five:*      Step forward into Kokutsu Dachi with right foot extended.

*Phase Six:*      With right hand holding Shuriken crossed at left hip, maintain Enzan No Metsuke.

*Phase Seven:*      Right hand starts to move in Nukitsuke throw, at same time move forward into Zenkutsu Dachi Kiai.

*Phase Eight:*      Maintain the follow-through. Inhale quietly.

*Phase Nine:*      Move hips back into Kokutsu Dachi. Take another Shuriken and repeat phases 6 through 8 until all Shuriken are thrown.

*Phase Ten:*      Step back from final Zenkutsu Dachi into Hachiji Dachi. Gaze Enzan No Metsuke. Zanshin mood.

*Phase Eleven:*      From Hachiji Dachi, adopt Heisoku Dachi. Hands at sides, execute Ritsurei.

The further Kata, to be included in a later volume, will cover movements from seiza, Iai Hiza and Iai Goshi. It will also contain various methods of special Samurai-style training.

# CHAPTER TEN

# Special Drills and "Secret Methods"

*"The river flows on and on, yet its
water is never the same"*

SHŌGUN JIZŌ

*Priest Kamochomei D.1216*

The true student of modern Shuriken Jutsu will surely wish to disown the small number of "bad apples" who are tending, by their rash actions to bring all who wish to improve their skill in Shuriken Jutsu, into disrepute. More than ever, safe usage of Shuriken and Shaken cannot be over-emphasised. The devastating power generated by expertly-thrown Shuriken attests to this. In competent hands, the arts of Shuriken Jutsu in the modern synthetic form, demand much of the student. Be warned they are hard task-masters. A carelessly thrown Shuriken or Shaken can spin off and return in a dangerous arc, severely injuring either the thrower or an innocent by-stander.

By their very nature, the drills mentioned here must be accomplished in a private area, which is totally enclosed. Spectators should be discouraged, but if there is no alternative, they should be seated in a position well behind the performer, behind a protective barrier which is constructed for that purpose and complies with the safety requirements of the area concerned – similar to that seen in a squash court. Chicken wire, no matter how fine the mesh is simply not good enough as a safety barrier.

It is vital that there is a target marshal, it is his responsibility to ensure absolute safety and discipline.

*Tool for removal of Shuriken.*

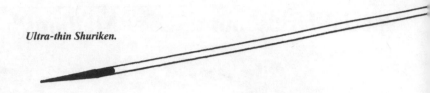

*Ultra-thin Shuriken.*

As with all modern sports, preparation and warm-up are an integral part of training. For, as all major trainers in the sports arena agree, the best performance and maximum training progress come from a performer whose body is sufficiently warmed up. I realise that many of you are already serious students; my words are directed at those of you who come fresh to this endeavour.

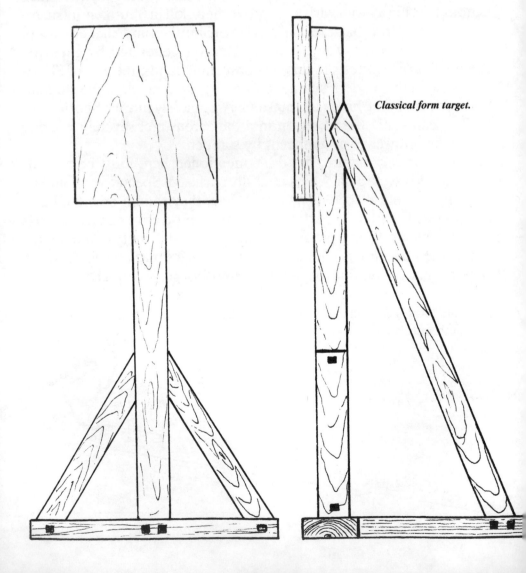

*Classical form target.*

## Special training drills:

*Exercise One:* The following is an endurance drill based upon aerobic principles, ie. encouraging the muscles to utilise an increased oxygen intake for maximum results and to do so safely.

*Note* such concepts as burn, wall, threshold of pain etc. are results of anaerobic exercise and have no place in this form of endurance training.

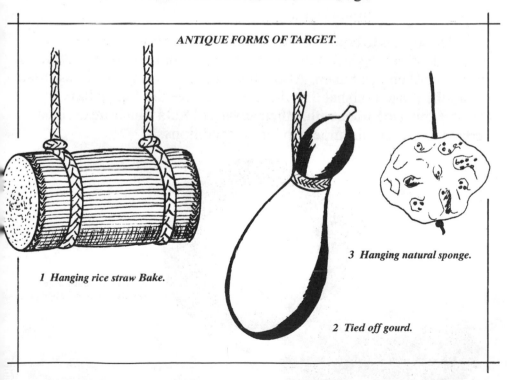

*ANTIQUE FORMS OF TARGET.*

*1 Hanging rice straw Bake.*

*3 Hanging natural sponge.*

*2 Tied off gourd.*

| | |
|---|---|
| *Phase One:* | Face the target area, with the Shuriken or Shaken laid out in front of you on a table or similar at waist height. |
| *Phase Two:* | Turn 180° and sprint a fixed distance. Turn and return. |
| *Phase Three:* | Face target in favoured throwing stance. |
| *Phase Four:* | Palm Shuriken and throw in quick succession. |
| *Phase Five:* | With index and middle fingers, take neck pulse (refer to chart). |
| *Phase Six:* | Retrieve Shuriken and repeat phase one through phase five; this time doubling the sprint distance. |
| **Comments:** | Observe and note percentage of accuracy. Rest until pulse drops by one third. Throw four more sets and note accuracy. |

The ideal is to equalize the two states; thus creating an accuracy independent of physical state. At no time during this exercise drill; run with Shuriken or Shaken. Also pause at the throwing line to ascertain from the range marshal that the area is clear – the Koryu have battle-field-proven methods within their advanced Kata which are designed to create consistent accuracy under all conditions.

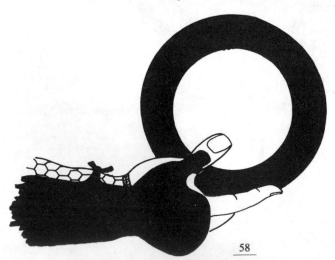

*Lead ring usually dropped on sentries to silence them.*

*Kokutsu fighting stance.*

*Jigotai (Squat Kiba Dachi) breathing posture.*

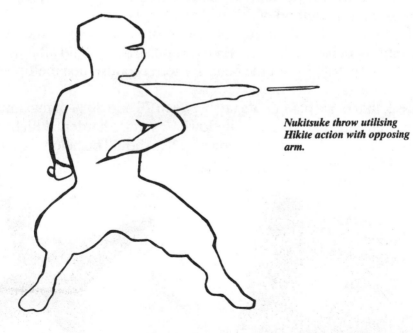

*Nukitsuke throw utilising Hikite action with opposing arm.*

*Exercise Two:*

*Phase One:*　　This exercise requires two targets set parallel to each other and a partner who calls out the action.

*Phase Two:*　　The performer faces the target squarely, a Shuriken in either hand.

*Phase Three:*　The partner then calls time. Throw both Shuriken, either left or right first.

*Phase Four:*　　The performer draws two more Shuriken and awaits the command to throw.

This is repeated any number of times, the purpose being to produce a balanced accuracy with left and right hand.

The modern synthetic tend to use the pulse-rate to gauge progress. The following is a method which will work:

With index and middle fingers, lightly press the side of the neck, where a strong pulse can be felt. Take this pulse before and after a ten-minute exercise drill. Count the beats for ten seconds, then multiply by six, to give beats per minute.

Check that beats per minute are between 70 and 80% of maximum for age – refer to chart. If beats are lower, exercise harder. If higher – slow down. Five minutes after exercise, pulse should be below 120, ten minutes after, below 100.

*Extra-large Shaken for use against horses.*

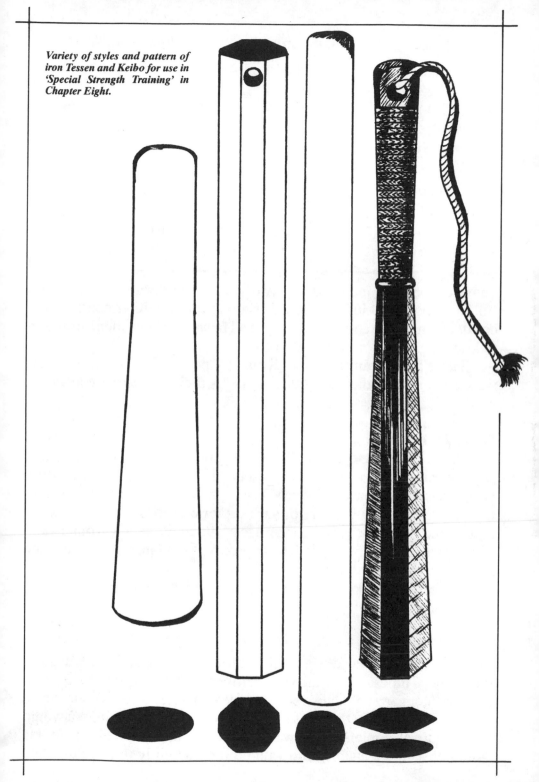

*Variety of styles and pattern of iron Tessen and Keibo for use in 'Special Strength Training' in Chapter Eight.*

| Age | Maximum Pulse | 70% − 80% |
|-----|---------------|-----------|
| 20 | 200 | 140 − 160 |
| 25 | 195 | 136 − 156 |
| 30 | 190 | 133 − 152 |
| 35 | 185 | 129 − 147 |
| 40 | 180 | 126 − 144 |
| 45 | 173 | 121 − 138 |
| 50 | 166 | 116 − 133 |
| 55 | 160 | 112 − 128 |
| 60 | 155 | 108 − 123 |
| 65 | 150 | 105 − 120 |

I have deliberately included only two forms of modern synthetic, which I offer as points of information. I reiterate that I neither condone nor condemn these practices; choosing in all fairness to give them an impartial hearing. It is you who must decide.

There are certain things which many persons, particularly in the west, have regarded as "secrets". I hope by their inclusion, to allay this myth.

## Shuriken Divination

Let me say first and foremost, that the techniques described here are, to say the very least, of dubious antiquity. However they are a colourful and rather quaint part of Japan's feudal past. I make no pretence as to the effectiveness or veracity of such esoteric exercises. It must be remembered that these grew up in an age of superstition, when omens and portents ruled both peasant and warlord. A man's fate was as sure as the strength of his sword arm.

Basically, Shuriken divination took two forms – neither, it must be said – secret.

## Method Number 1

On the battlefield, usually among the lower orders, such as Ashigaru, the Kanji (Chinese ideogram) for victory and defeat would be written, or if the Ashigaru could not read; a circle would represent victory and a cross, defeat. Setting the signs up on a suitable rest the Ashigaru would take whatever implement he was considering using as a Shuriken. He would pray to his divinity and with both eyes tightly shut, would fling the Shuriken at the target. On hearing the concussion, he would open one eye and warily view the target. If the Shuriken had speared the symbol for defeat, he would retrieve the Shuriken and throw again. Being no fool, he would continue throwing until victory was finally speared. He would then open both eyes, and confident of victory, prepare for battle. The concept of one eye/two eye, is quite popular. Even today, one can purchase, at shrine festivals, a papier maché doll of the Buddhist patriarch, Daruma, which comes with no eyes, the idea being, that with a brush, one dots in an eye and makes a wish. When the wish is fulfilled, one dots the other eye to "seal" the fate. Nothing, in truth, is new.

## Method Number 2
This second method is almost the Japanese version of English Victorian parlour tricks. It would take place in the household of a middle-rank Samurai. On an auspicious day, a piece of Hosho paper containing a circle divided into twelve Zodiacal signs of:-

| Ne | – | Rat | Uma | – | Horse |
|-----|----|--------|---------|----|---------|
| Ushi | – | Ox | Hitsuji | – | Sheep |
| Tora | – | Tiger | Saru | – | Monkey |
| U | – | Hare | Tori | – | Chicken |
| Tatsu | – | Dragon | Inu | – | Dog |
| Mi | – | Snake | I | – | Hog |

*Front view.*

*Side View.*

*Position for covert securing of Shuriken in Obi – note that Shuriken is well away from vertebrae in case of Ukemi.*

*Complete.*

And the four cardinal points of –

**Seirio** – *the green dragon guardian of the east.*
**Gembu** – *the dark warrior guardian of the north.*
**Shujaku** – *the red bird guardian of the south.*
**Biakko** – *the white tiger guardian of the west.*

To the circle was then added *San Sai*, the three powers of nature, or *Sanze* the three states of existence.

The players then requested the answer to a personal question or problem and threw a most ornate form of lightweight Shuriken or Kodzuka.

Again, I cannot really comment on the veracity of these methods, merely inserting them as an item of entertainment, on a par with the Genji incense game.

So, my friends, we have come a long way in a short time on our voyage of discovery of the enigma that is Shuriken Jutsu. Outside the dawn light is over-powering the flicker of my candle. It is time to call a halt, I would however, like to leave you with this thought:–

> *"As we stand on the threshold of the 21st century, we must truly realise, we are neither Samurai nor Ninja, but we are the guardians of their heritage"*

**Katsumi Toda**

# Bibliography

For those of you who wish to pursue their studies with a little more depth, I include the following bibliography, which I trust will be of assistance. For those of you who do not have access to Japanese sources, I also include works in European languages.

### Japanese language

| | | |
|---|---|---|
| Hagio Takashi | Nihon Kendo To Token | Tokyo 1943 |
| Hakuseki Arai | Honcho Gunkiko | Private collection |
| Imaeda Aishin | Zenshu No Rekishi | Tokyo 1962 |
| Imamura Yoshio | Nihon Budo Zenshu 7 volumes | Tokyo 1966 |
| Miyamoto Musashi | Gorin No Shō | NBZ Vol Tokyo 1961 |
| Nagayoshi Saburo | Nihon Bushido Shi | Tokyo 1932 |
| Okada Akio | Kama Kura Bushi | NNR Vol 4 Tokyo 1963 |
| Omori Sogen | Ken to Zen | Tokyo 1958 |
| Sato Kan'Ichi | Nihon To Token | Tokyo 1963 |
| Sugino Yosio | Tenshin Shoden Katori Shinto Ryu Budo Kyohan | Tokyo 1936 |
| Tsuka Hara Bokuden | Bokuden Ikun Sho | NBZ Vol 2 Tokyo 1961 |
| Yamada Jirokuchi | Nihon Kendo Shi | Tokyo 1960 |

## European Language

| | |
|---|---|
| Draeger Donn F. | Classical Budo |
| | Classical Bujutsu |
| | Modern Bujutsu and Budo |
| | Ninjutsu, the art of invisibility |
| With Warner Gordon | Japanese Swordsmanship |
| Hütterott G. | Das Japanische Schwert |
| Kammer Reinhard | Tengu Geijutsu Ron |
| Knutsen Roald | Japanese Polearms |
| Robinson B.W. | The Arts of the Japanese Sword |
| | Primer of Swords |
| | Arms and Armour of Old Japan |
| Robinson H.R. | Armour book in Honcho Gunkiko |
| | Japanese Arms and Armour |
| Turnbull S.R. | Samurai – A Military History |
| | Book of the Samurai |
| Warner G. with Sasamori J. | This is Kendo |

## Other Titles published by Dragon Books

Nunchaku Dynamic Training
*By Hirokazu Kanazawa 8th Dan*

Shadow of the Ninja
*By Katsumi Toda*

Revenge of the Shogun's Ninja
*By Katsumi Toda*

Advanced Shotokan Kata Series
*By Keinosuke Enoeda 8th Dan*

Volume 1
Bassai-Dai : Kanku-Dai : Jion : Empi : Hangetsu

Volume 2
Bassai-Sho : Kanku-Sho : Jiin : Gankaku : Sochin

Volume 3
Tekki-Nidan : Tekki-Sandan (2 versions) : Nijushiho : Gojushiho-Dai
: Gojushiho-sho

Dynamic Kicking Method
*By Masafumi Shiomitsu 7th Dan*

The Kubotan Keychain
*By Takayuki Kubota*

Balisong – Iron Butterfly
*By Cacoy Hernandez*

Ninja Death Vow
*By Katsumi Toda*

When the Going Gets Tough
*By Col. M. Smythe*

Naked Blade – A Manual of Samurai Swordsmanship
*By Toshishiro Obata 7th Dan*

Dynamic Power of Karate
*By Hirokazu Kanazawa 8th Dan*
*(first published as Kanazawa's Karate)*

Ninja Sword-Art of Silent Kenjutsu
*By Katsumi Toda*

Ninja Training Manual – A Treasury of Techniques
*By Yukishiro Sanada*

**Dragon Books** are available from branches of B. Dalton Booksellers, Walden Books and all good martial arts and general bookstores. If you have difficulty obtaining any of these titles, please contact the publisher direct. Orders under $10 can be filled for the advertised price plus $1.50. For orders over $10 simply add 10% to the value of your order to cover freight and handling charges. Overseas customers, please contact us for details of export shipping costs.

**Dragon Books P.O. Box 6039 Thousand Oaks CA 91359 USA**

**U.K. Distributor: Sakura Publications P.O. Box 18, Ashtead Surrey KT21 2JD**

*Phototypeset in the United Kingdom by Concise Graphics Ltd. Hammersmith London*